Progress with Oxford

Age 5-6

Numbers and Counting up to 100

Hello! I'm Tally.

Contents

Key

 Draw

 Write

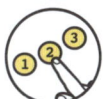 Count

 Match

 Colour

 Trace with pencil

 Play together

 Circle

 Find the sticker

OXFORD
UNIVERSITY PRESS

Read and write numbers 0–10

Trace the numbers and number words.

Write the word on the line.

0	zero	zero
1	one	
2	two	
3	three	
4	four	
5	five	
6	six	
7	seven	
8	eight	
9	nine	
10	ten	

✎ Write the missing numbers.

Remember to use the number line at the top and bottom of the pages to help you.

Well done!

🏐 Play with numbers 0 to 10.

Take turns to challenge a friend to count some objects (cars, people, trees, toys or books) between 1 and 10. Then find the number written down somewhere (for example, in a book, on a clock, a house number, a bus number).

Be a number word detective! Look for number words around your home.

Give yourself a sticker

Now – track how you're doing on page 32!

Read and write numbers 11–20

 Trace the numbers and number words.

 Write the words on the line.

Check back to help you remember the words.

11	eleven	
12	twelve	
13	thirteen	
14	fourteen	
15	fifteen	
16	sixteen	
17	seventeen	
18	eighteen	
19	nineteen	
20	twenty	

 Draw dots to match each number.

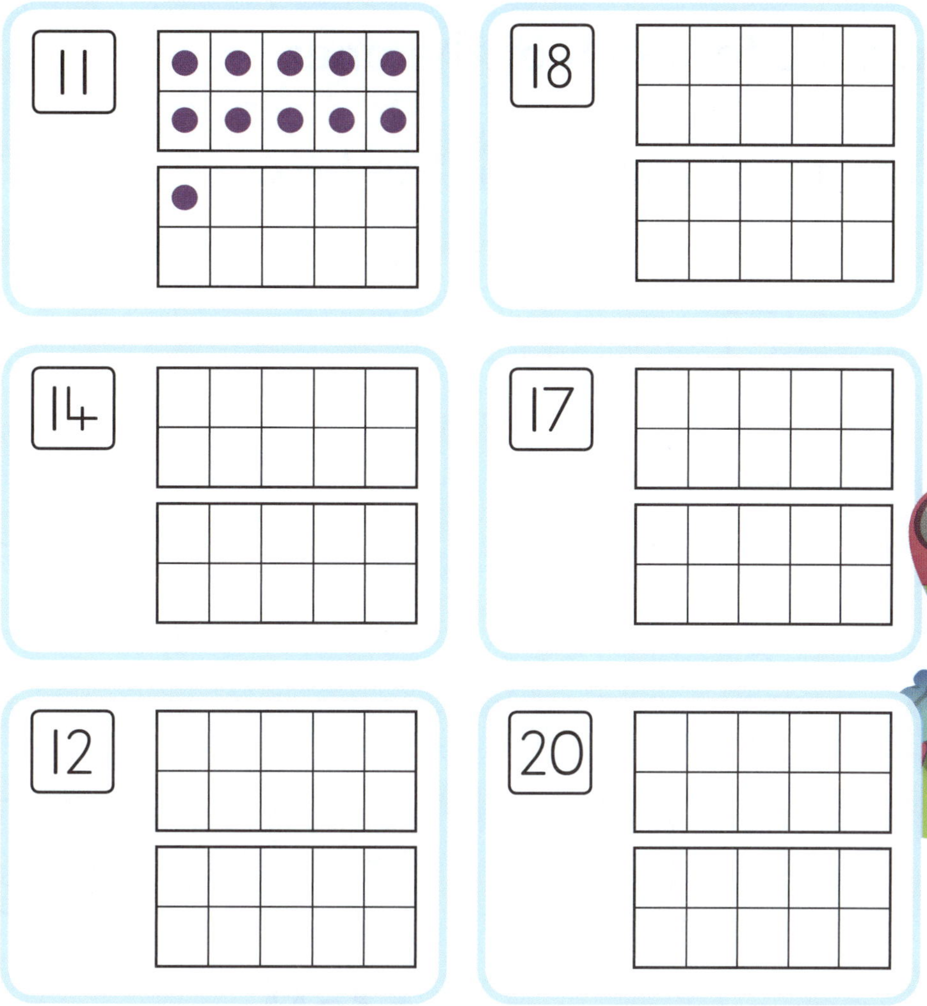

11	● ● ● ● ● / ● ● ● ● ● / ●
18	
14	
17	
12	
20	

What can you do next?

Well done!

Give yourself a sticker

 Play with numbers 11 to 20.

Use chalk on the ground outdoors (or masking tape indoors) to draw a number line from 11 to 20.

Play *Snap* with a friend. Write the numbers 11 to 20 and the words 'eleven' to 'twenty' on small pieces of paper. Shuffle the 'sets' separately and turn over the top card from each pile. Take it in turns to turn over a card from each pile. Call *Snap!* if the number and word match.

Now – track how you're doing on page 32!

Tens and ones up to 20

 Draw a cube for each of the ones.

 Write the total number in the box.

10 4 **14**	10 1
10 6	10 3
10 5	
10 9	

Well done!

Give yourself a sticker

Now – track how you're doing on page 32!

Count and write numbers 0–100

 Write in the missing numbers.

1				5	6	7	8	9	10
11	12	13	14	15	16		18	19	20
21	22	23	24	25	26	27	28	29	
	32	33	34			37	38	39	40
41	42	43	44	45	46	47	48		
	52	53	54	55	56	57	58	59	60
61	62		64	65	66	67	68	69	70
71	72	73				77	78	79	80
	82	83	84	85	86	87		89	90
		93	94	95	96	97	98	99	

Use the number line at the top and bottom of the page to help you.

 Play with numbers 0 to 100.

Be an outdoor number explorer! Pick a number between 1 and 100. Look for that number when you are out and about. Choose a different number next time you go out.

Now – track how you're doing on page 32!

Count forwards 0–100

 Count forwards.

 Find the stickers to fill in the missing number balloons.

 Write in the missing numbers on the flags.

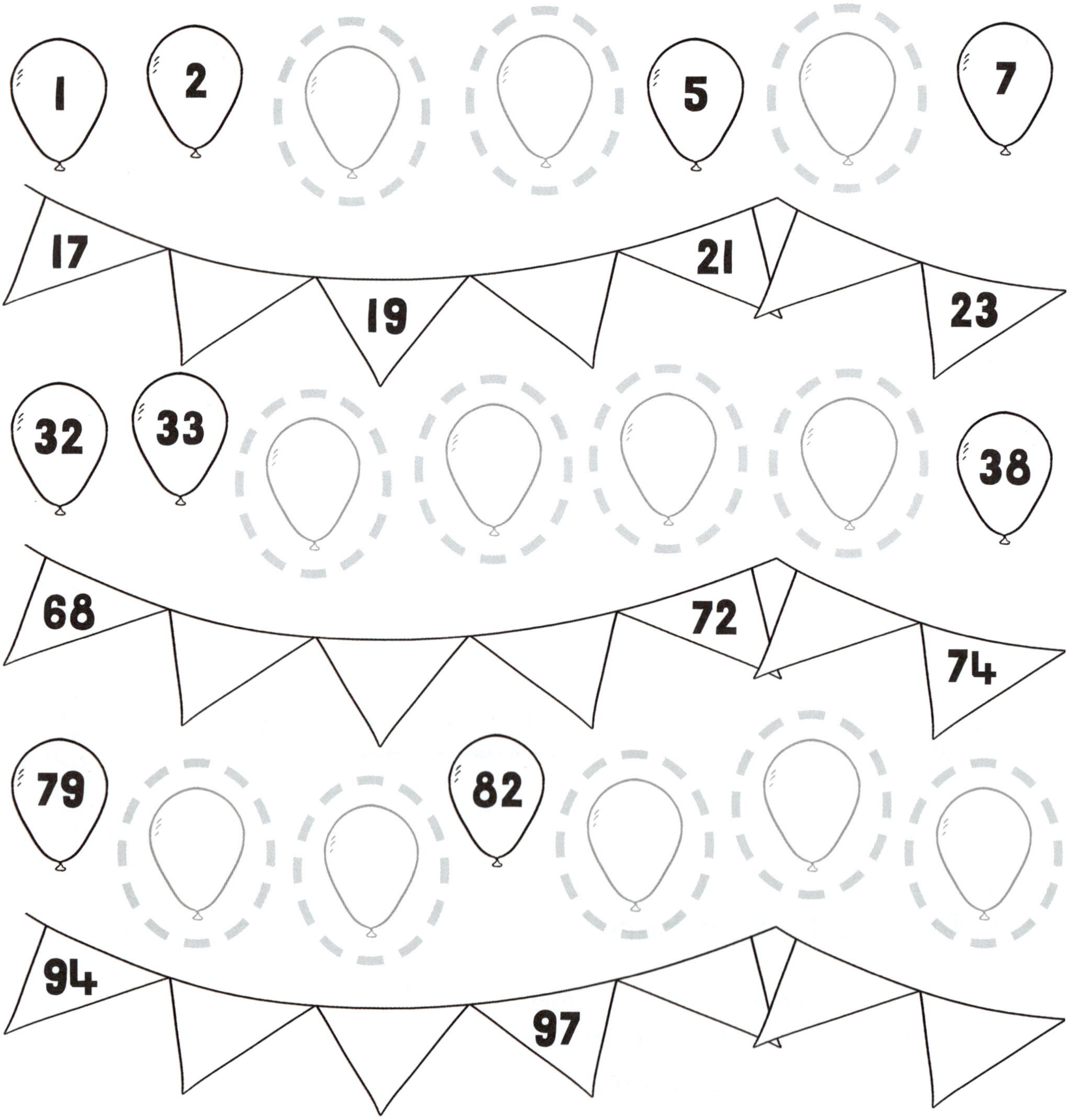

1   2   5   7

17   19   21   23

32   33   38

68   72   74

79   82

94   97

Count backwards 0–100

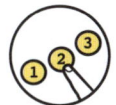 Count backwards.

 Write the missing numbers in the gaps.

5 ___ ___ ___ 2 ___ 0

13 ___ ___ ___ ___ 9 8

28 27 ___ ___ ___ ___ 23

41 ___ ___ ___ 37 ___ ___

72 ___ ___ ___ 68 ___ ___

95 ___ ___ ___ ___ ___ 90

What animal chases a caterpillar? A dogerpillar!

Give yourself a sticker

Now – track how you're doing on page 32!

Count in 2s

Count in 2s using the pairs of eyes to help you.

Write the missing numbers in the boxes.

| 0 | | 4 | | | 10 |

How many flip-flops are there? Count in 2s.

Write your answer in the sandcastle.

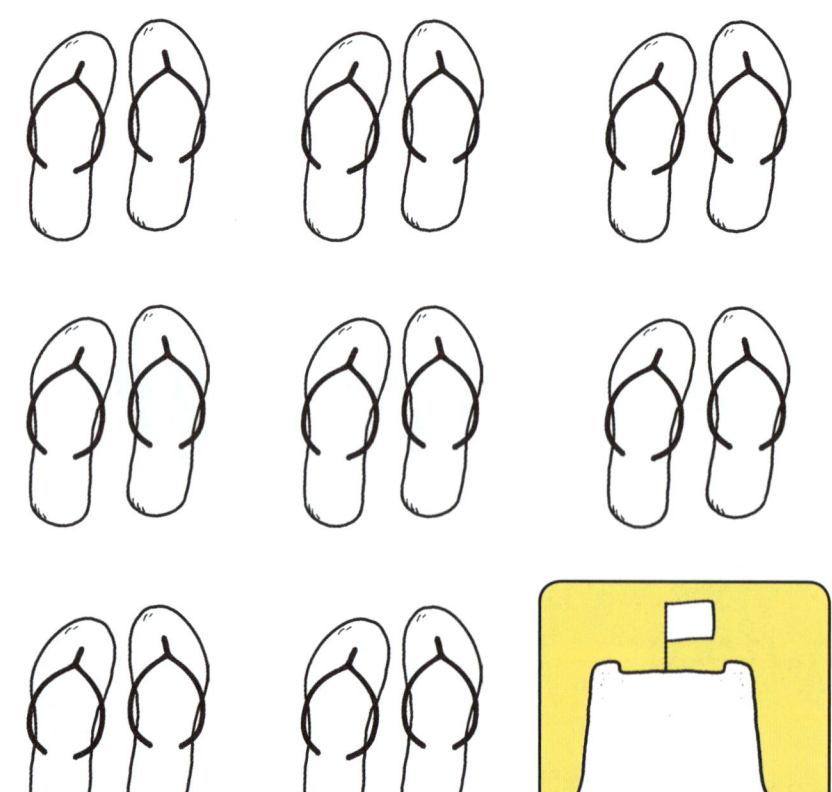

16

 Count up in 2s and colour the squares to help Tally find the ice cream.

		0	2	3	20	18
		1	4	9	14	15
9	7	8	6	10	9	16
14	13	10	5	18	20	
12	11	12	14	16	17	

Play with counting in 2s.

Jump up and down using both feet and count in 2s each time you jump. You could jump on the spot, jump in a line or jump all around!

Next time you are outside, count the number of wheels you see on bicycles. Count two for each bicycle you see.

Use the number line if you need help to count in 2s.

Give yourself a sticker

Now – track how you're doing on page 32!

Count in 10s

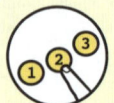 Count in 10s using the toes to help you.

 Write the missing numbers in the boxes.

| 0 | | | 30 | | |

 Start at 10. Count in 10s.
Colour every 10th number.

1	2	3	4	5	6	7	8	9	10
11	12	13	14	15	16	17	18	19	20
21	22	23	24	25	26	27	28	29	30
31	32	33	34	35	36	37	38	39	40
41	42	43	44	45	46	47	48	49	50
51	52	53	54	55	56	57	58	59	60
61	62	63	64	65	66	67	68	69	70
71	72	73	74	75	76	77	78	79	80
81	82	83	84	85	86	87	88	89	90
91	92	93	94	95	96	97	98	99	100

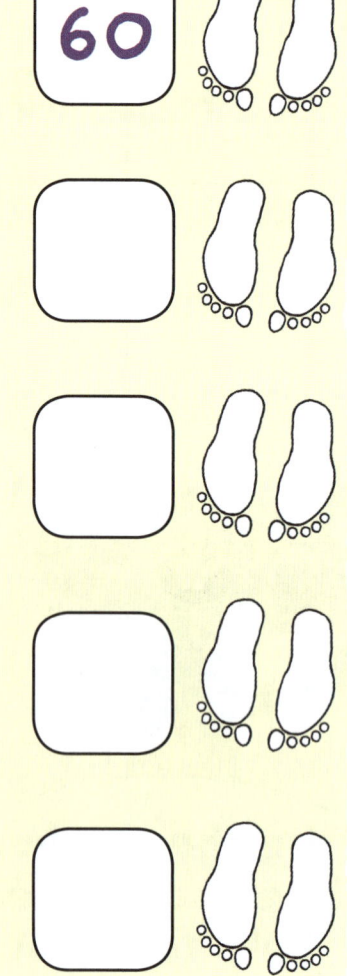

60

 Count in 10s to find the 10s numbers.

 Circle them to turn them into basketballs.

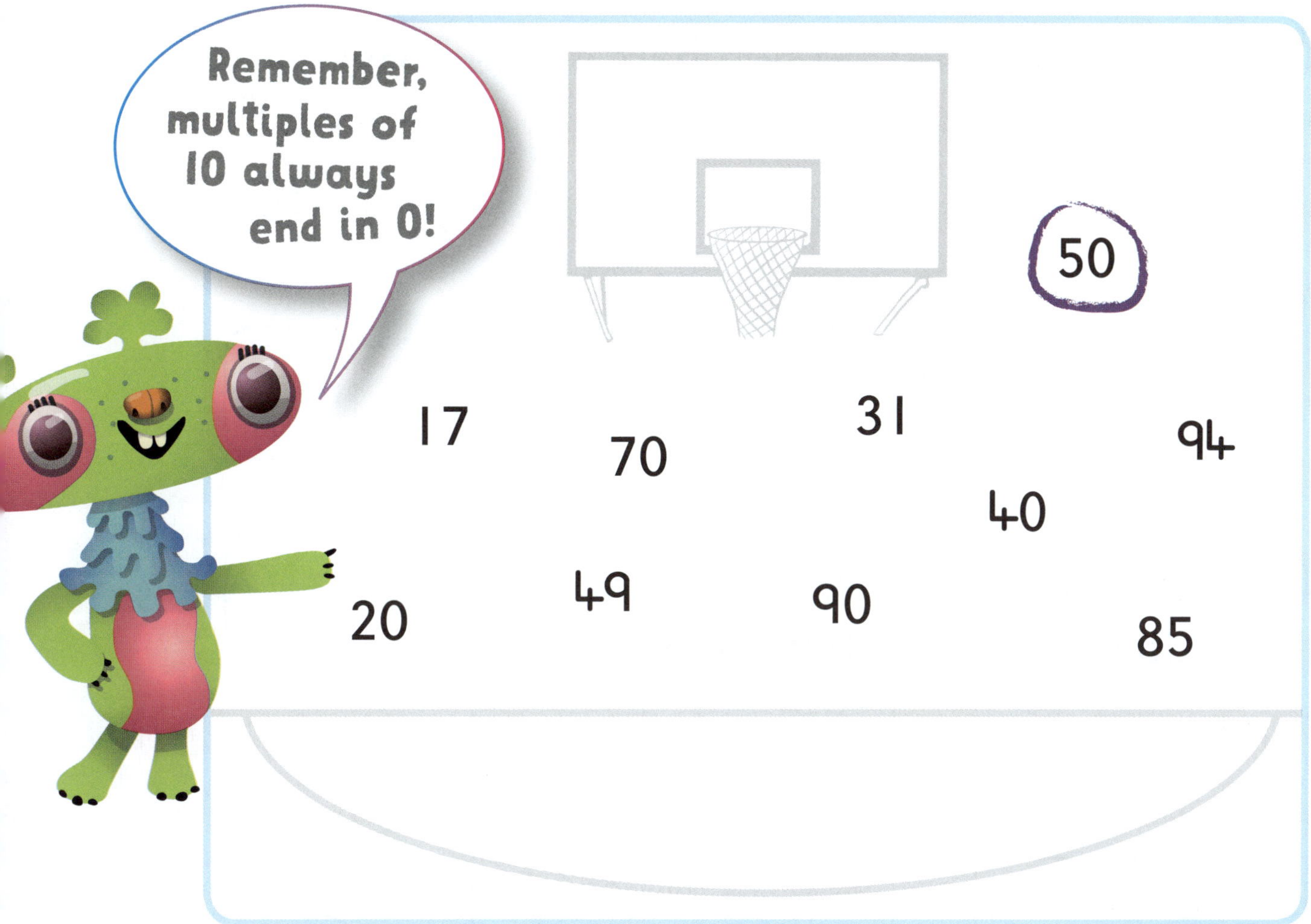

Remember, multiples of 10 always end in 0!

50

17

70

31

94

40

49

90

85

20

 Play with counting in 10s.

Close and open both hands at the same time to show ten fingers. Start at zero and each time you open your hands, count ten more.

Get different numbers of your friends and family together. Ask them to put their hands, feet or both out in front of them. Count in 10s to see how many fingers and toes there are altogether.

Give yourself a sticker

Now – track how you're doing on page 32!

Count in 5s

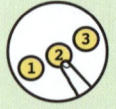 Count in 5s using the fingers to help you.

 Write the missing numbers in the boxes.

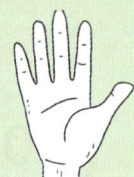 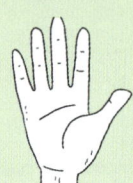 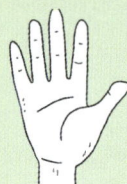 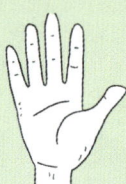 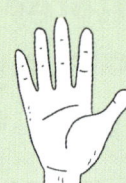

| 0 | | | | 20 | |

 Start at 5. Count in 5s.
Colour every 5th number.

1	2	3	4	5	6	7	8	9	10
11	12	13	14	15	16	17	18	19	20
21	22	23	24	25	26	27	28	29	30
31	32	33	34	35	36	37	38	39	40
41	42	43	44	45	46	47	48	49	50
51	52	53	54	55	56	57	58	59	60
61	62	63	64	65	66	67	68	69	70
71	72	73	74	75	76	77	78	79	80
81	82	83	84	85	86	87	88	89	90
91	92	93	94	95	96	97	98	99	100

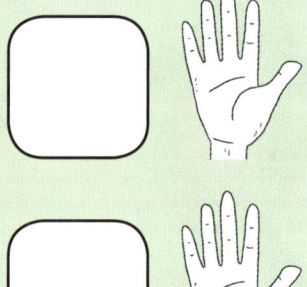

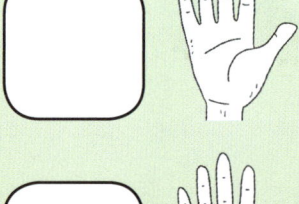

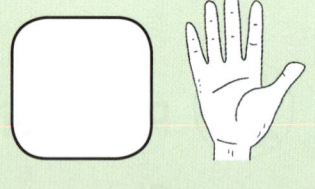

45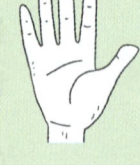

 Count in 5s from 0 and join the dots to complete the picture.

5
0
100
10
95
15
90
20
80   85   25   30
75   35
70   40
55
65   45
60   50

Tap your fingers on the table as you count.

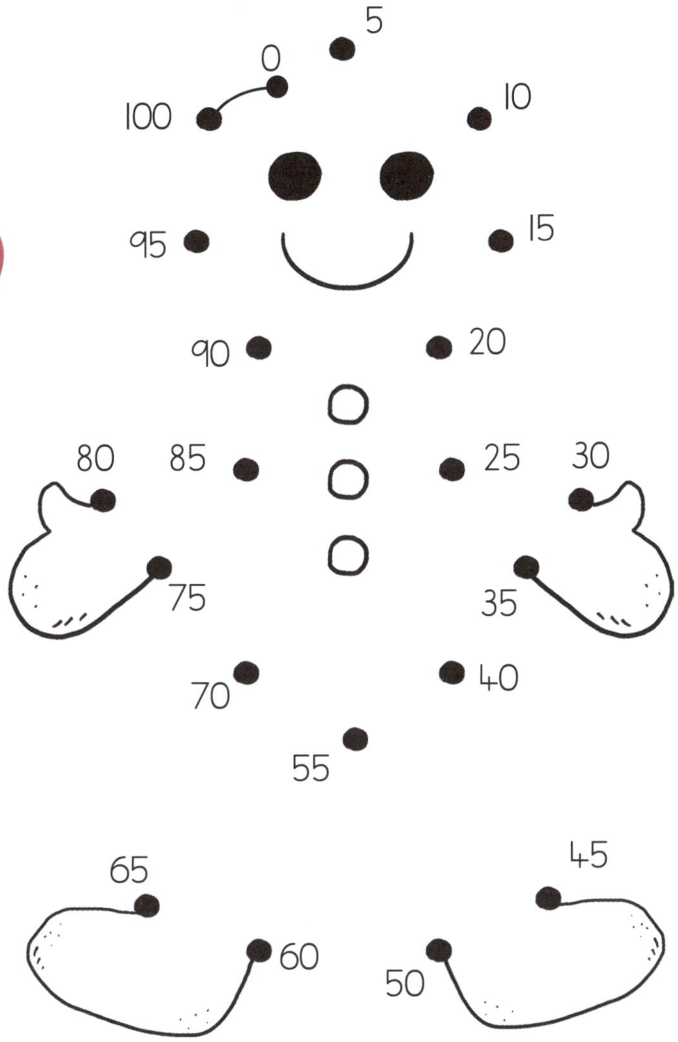

 **Play with counting in 5s.**

Close and open one hand to show five fingers. Start at zero and each time you open your hand count five more.

Be an outdoor number explorer! Look for as many multiple of 5 numbers as you can. Check house numbers, bus numbers, advertisements, car number plates, clocks, signposts and any other places you can think of!

Give yourself a sticker

Now – track how you're doing on page 32!

Number puzzles

 Write each number in words to fill in the crossword.

five   seven

zero   nine

four

eight

six

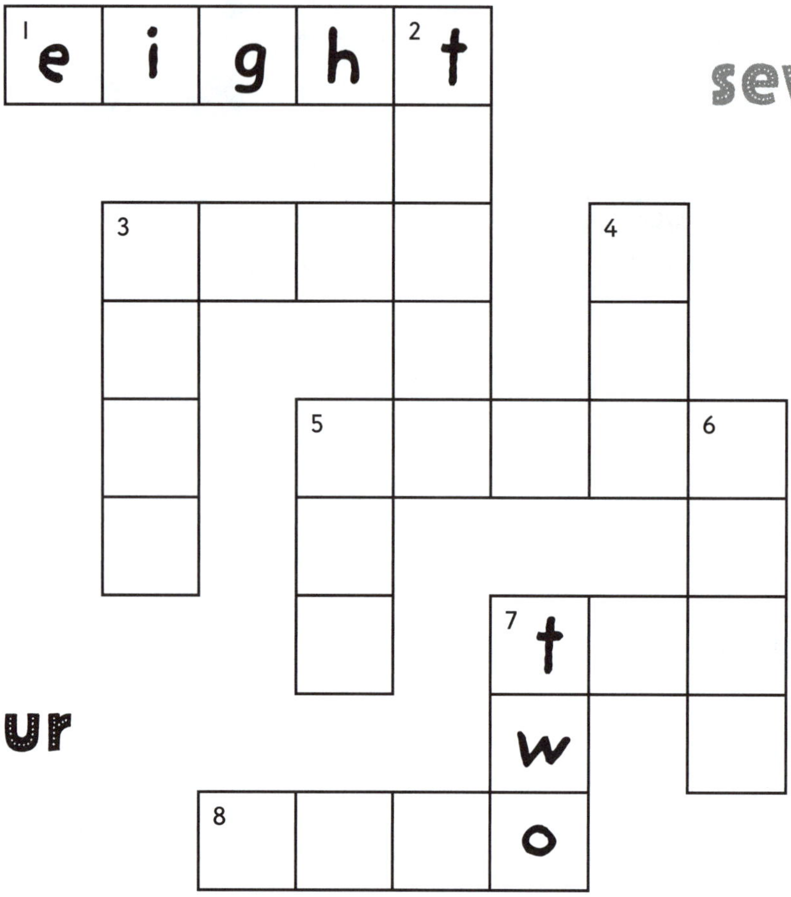

Across
1. 8
3. 4
5. 7
7. 10
8. 0

Down
2. 3
3. 5
4. 1
5. 6
6. 9
7. 2

three   two

one   ten

Crossword grid:
- 1 Across: e i g h t
- 7: t / w / o

Stickers for page 8

Stickers for page 24

Stickers for page 26

Stickers for pages 30–31

Reward Stickers

Stickers for pages 30–31

Character stickers

 ★★ Circle each of the number words in the wordsearch.

```
s  i  x  t  e  e  n  d  f
e  i  g  h  t  e  e  n  o
v  e  p  i  a  i  t  f  u
e  l  e  r  s  t  w  i  r
n  e  n  t  y  w  e  f  t
t  v  g  e  a  e  l  t  e
e  e  u  e  n  n  v  e  e
e  n  i  n  e  t  e  e  n
n  d  n  e  d  y  d  n  i
```

eleven   sixteen

twelve   seventeen

thirteen   eighteen

fourteen   nineteen

fifteen   twenty

Where is the bonus animal word?

Give yourself a sticker

Now – track how you're doing on page 32!

Number lines 0–100

 Write the missing numbers.

0   1   2   ☐   4   5   ☐   7   ☐   9   ☐

20   ☐   22   ☐   24   25   26   27   ☐   29   ☐

☐   41   42   43   ☐   45   ☐   47   48   49   ☐

☐   91   92   93   ☐   ☐   96   ☐   98   99   100

 Write each number in the correct place on the number line.

Look for the smallest number and write it in the first box.

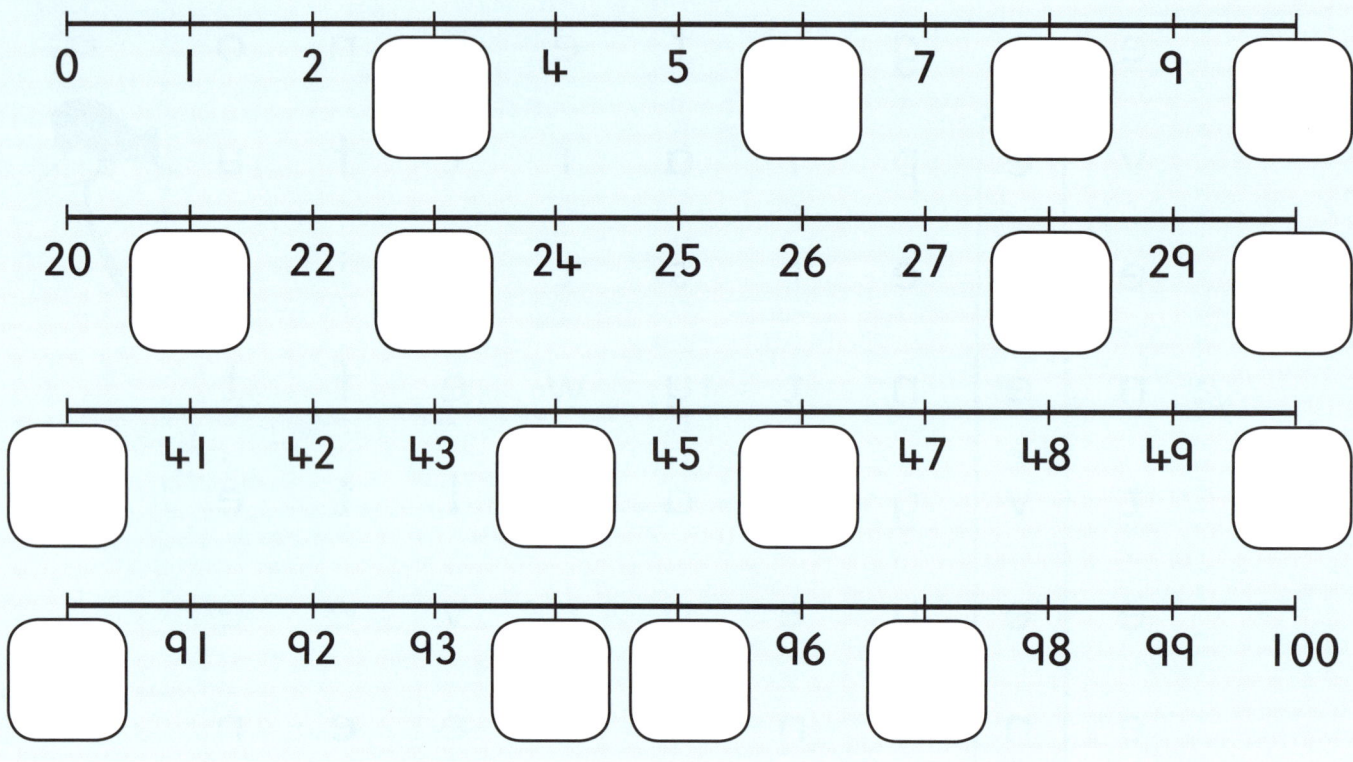

58   54   56   50   57   52   53   59   55   51

☐ ☐ ☐ ☐ ☐ ☐ ☐ ☐ ☐ ☐

Estimation

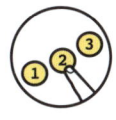 Estimate how many animals there are and then count them.

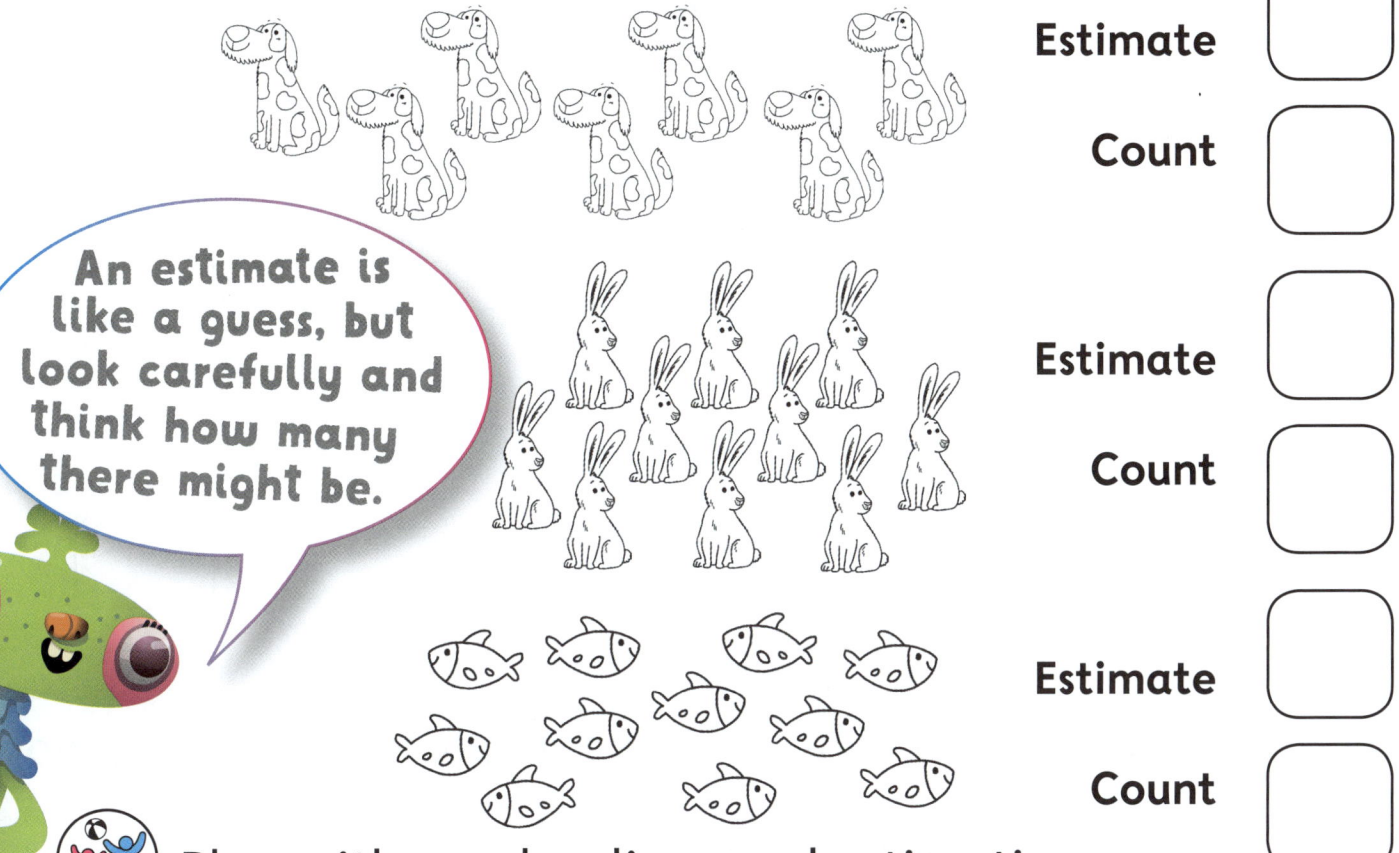

Estimate

Count

Estimate

Count

Estimate

Count

> An estimate is like a guess, but look carefully and think how many there might be.

Play with number lines and estimation.

Use string, paper and clothes pegs to make a mini washing line. Write some consecutive numbers on small pieces of paper (such as 0 to 10, 20 to 30 or 60 to 70). Shuffle them and see how quickly you can make a number line by pegging the numbers onto the washing line in the correct order.

Put 20 small items, like marbles or pieces of dried pasta, in a small bag or box. Put a handful of them on the table. Estimate and then count how many you took out.

Give yourself a sticker

Now – track how you're doing on page 32!

One more and one less

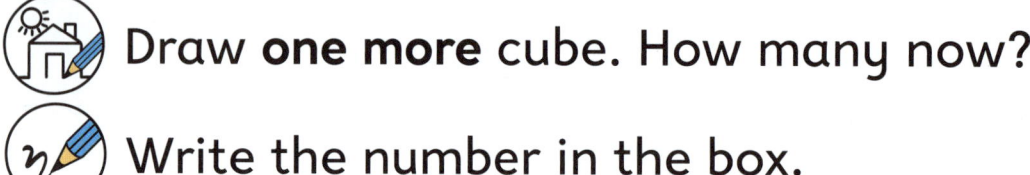

Draw **one more** cube. How many now?

Write the number in the box.

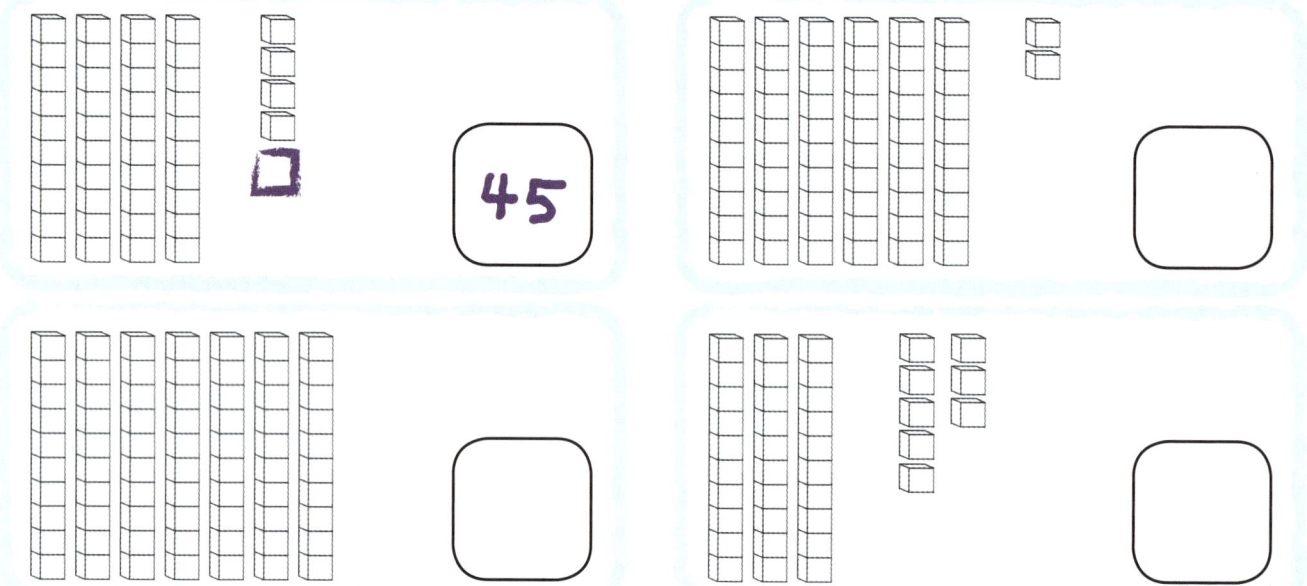

45

Cross out one cube to make **one less**. How many now?

Write the number in the box.

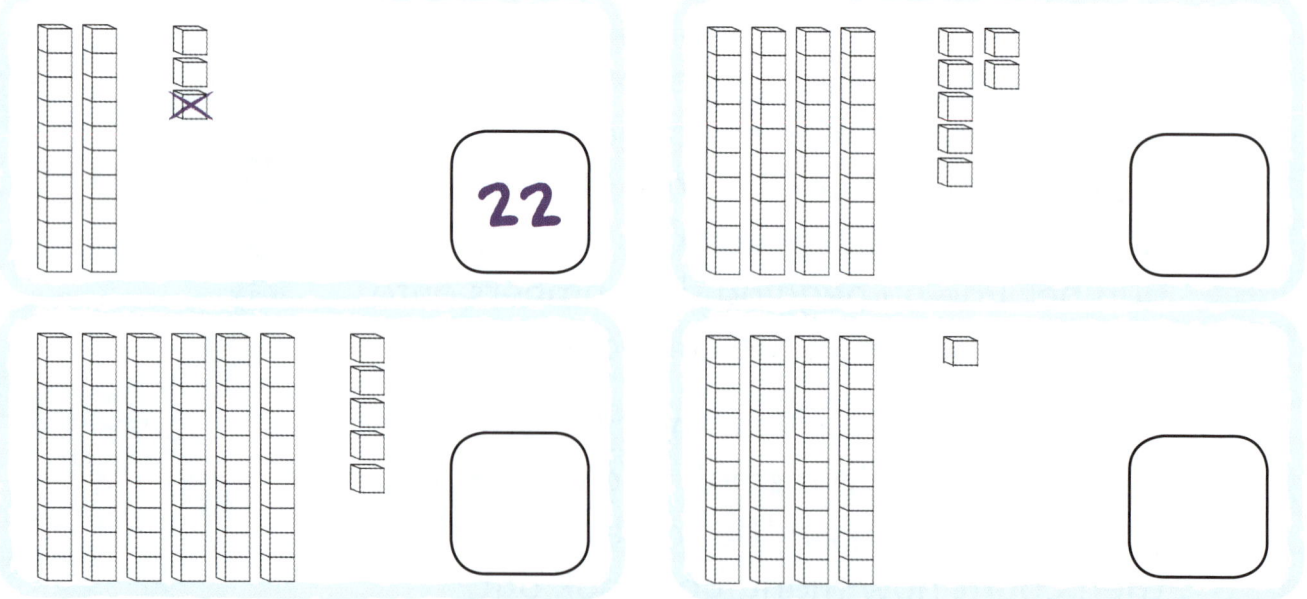

22

 Write one less and one more on the wings.

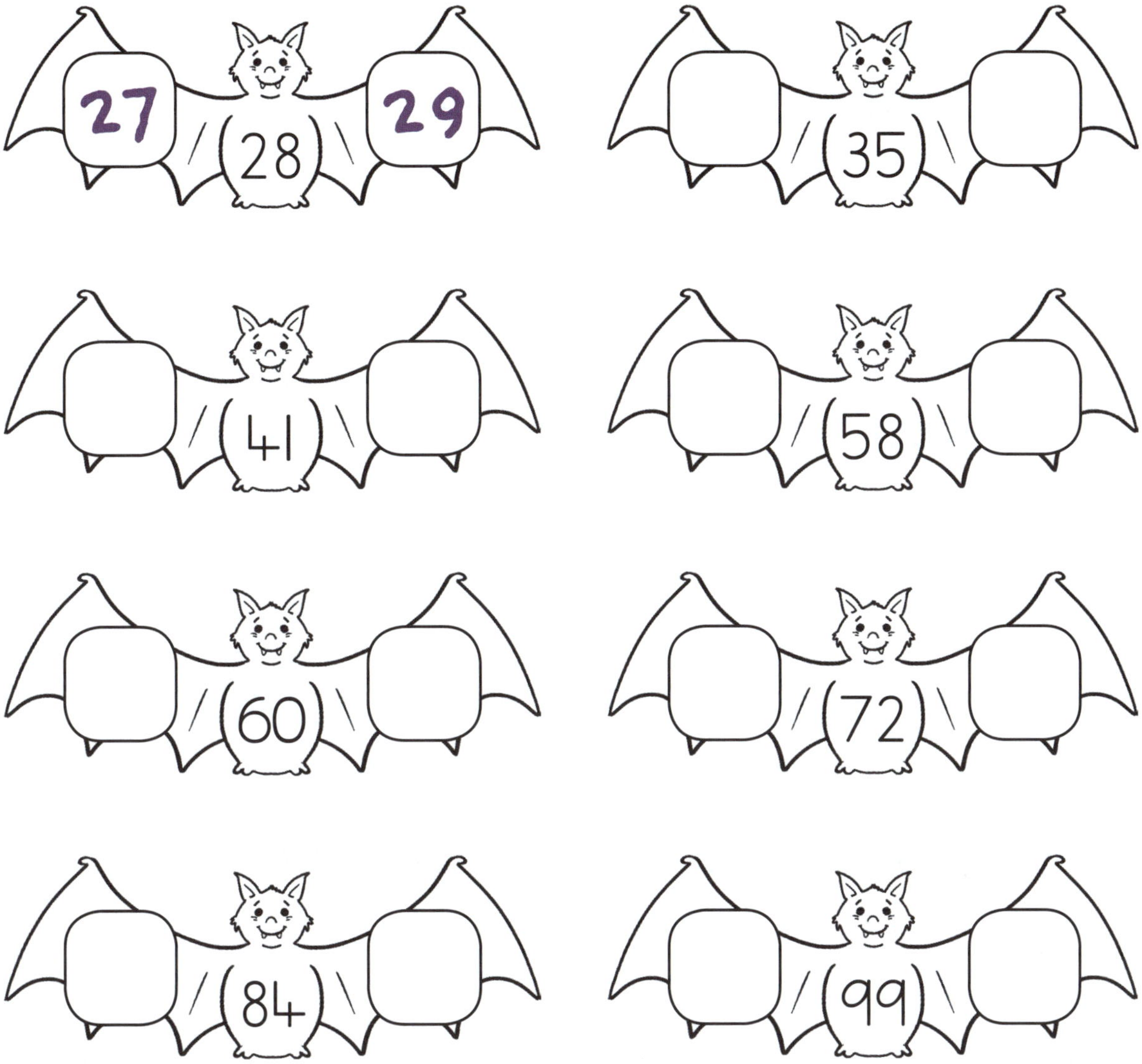

27 (28) 29

[ ] (35) [ ]

[ ] (41) [ ]

[ ] (58) [ ]

[ ] (60) [ ]

[ ] (72) [ ]

[ ] (84) [ ]

[ ] (99) [ ]

Count backwards to find one less. Count forwards to find one more.

Give yourself a sticker

Now – track how you're doing on page 32!

0 1 2 3 4 5 6 7 8 9 10 11 12 13 14 15 16 17 18 19 20 21 22 23 24 25

More than and less than

Circle the number in each pair that is **more than** the other number.

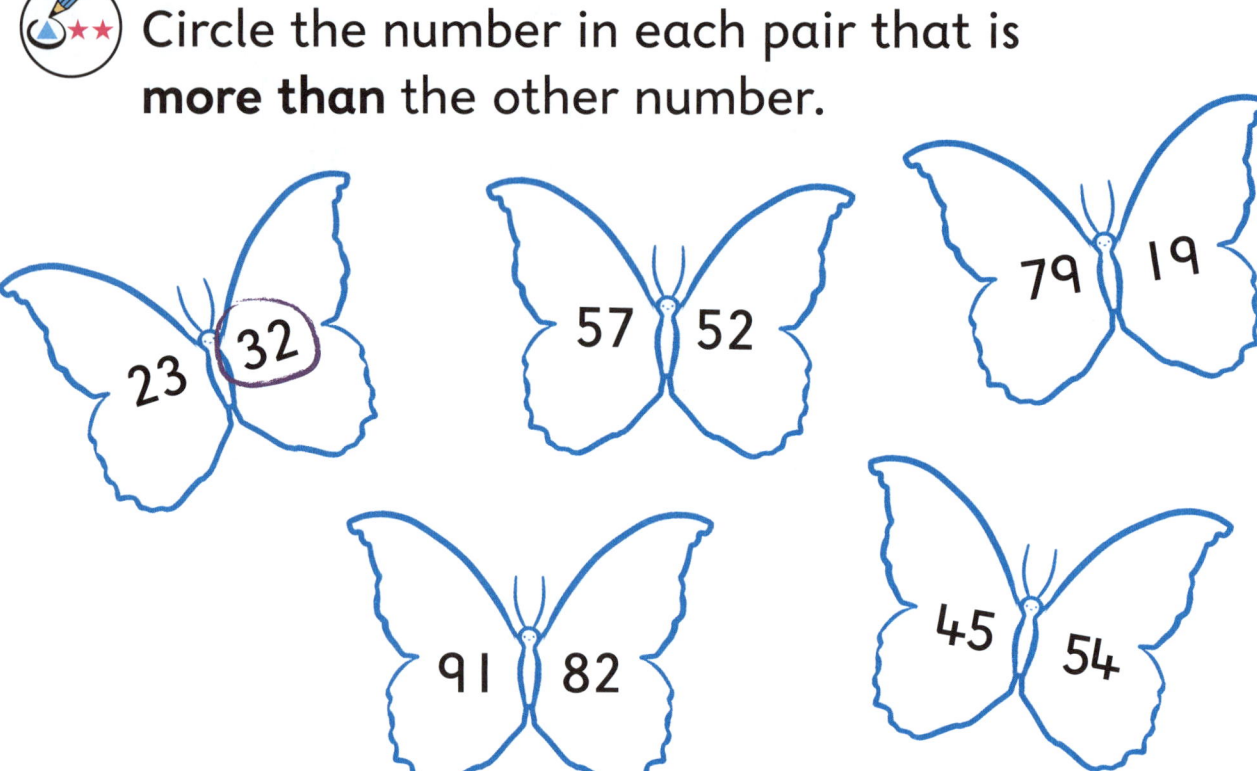

Circle the number in each pair that is **less than** the other number.

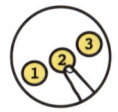 Count the number of balls.

 Draw bats so there are **more than** the number of balls.

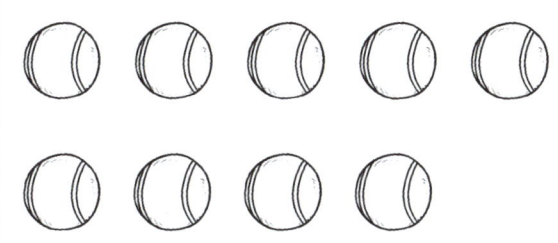

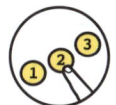 Count the number of watering cans.

 Draw flowers so there are **less than** the number of watering cans.

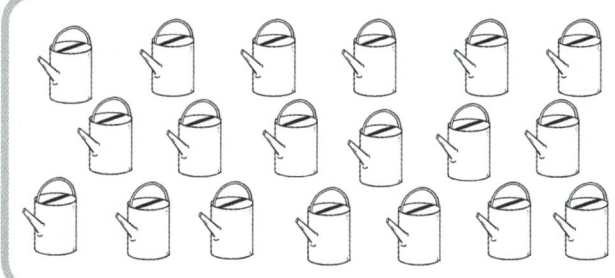

 Play with more than and less than.

You will need some coins. Pick up two different coins and look for the number on them. Which number is more than the other? Try different pairs of coins. Work out which one is less than the other, too.

Challenge a friend or friends to a hopping competition. Who hops more than the other(s)? Who hops less than the other(s)?

Give yourself a sticker

Now – track how you're doing on page 32!

In between

 Find the sticker with the number that comes in between.

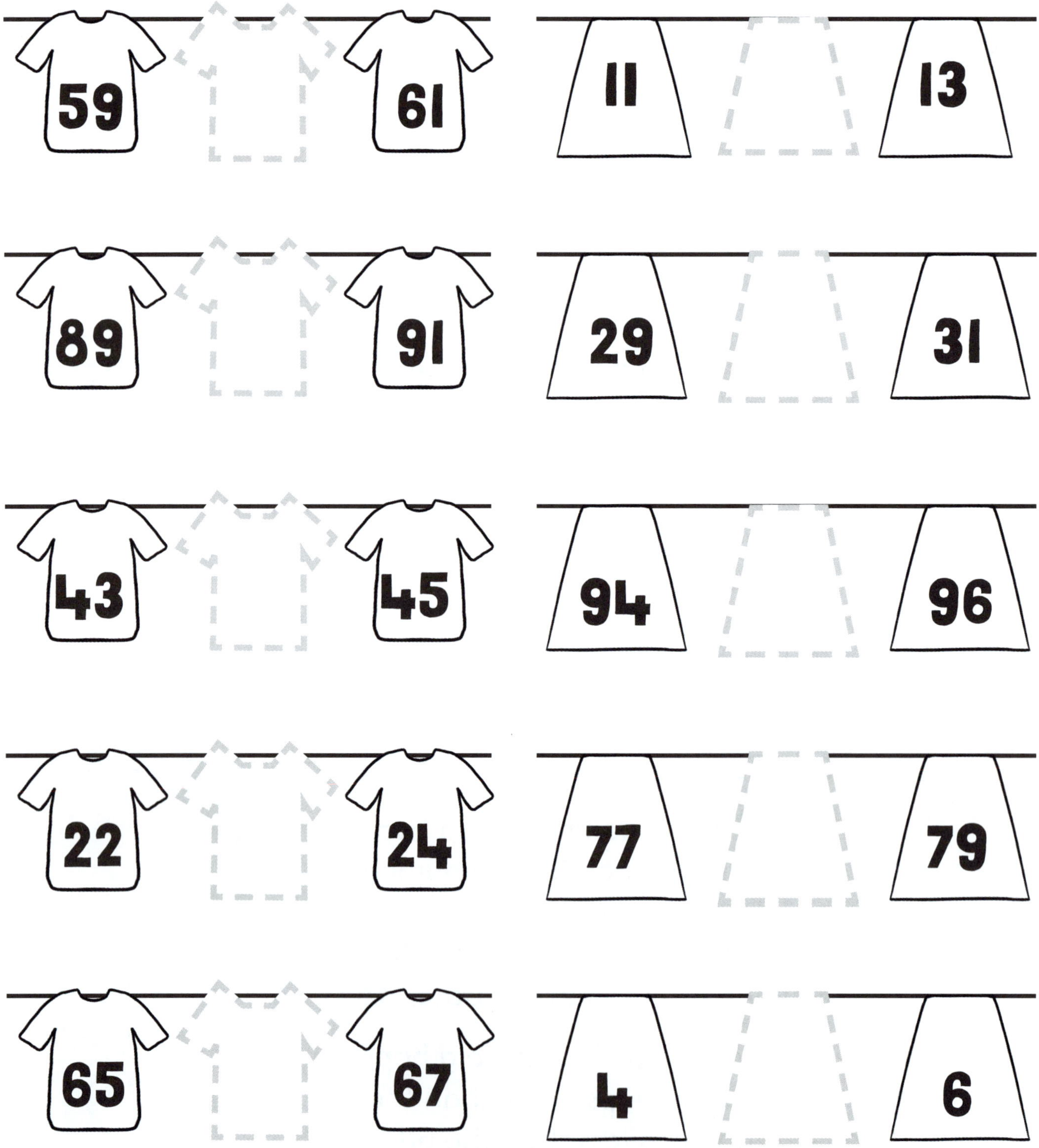

59   61   11   13

89   91   29   31

43   45   94   96

22   24   77   79

65   67   4   6

Equal to

Let's do more counting!

 Draw lines to match pairs of equal numbers.

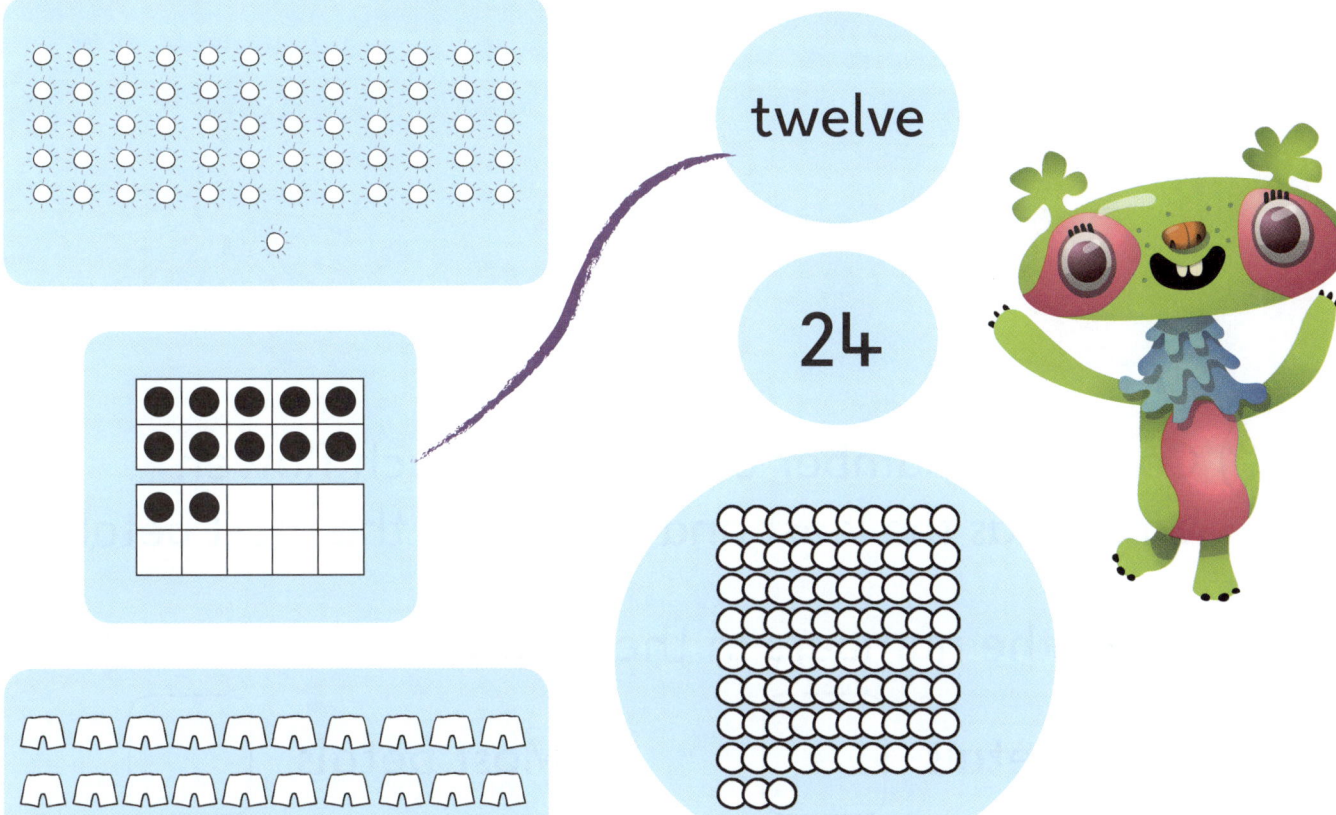

twelve

24

83

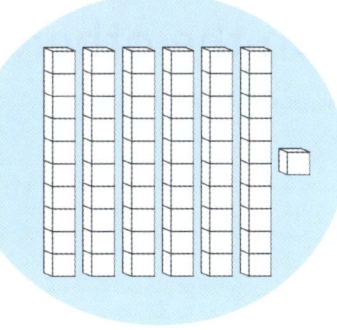 Play with in between and equal to.

Use as many different ways as you can to show the number 15. Write, draw, find objects – be creative! Choose a different number and have another go.

Give yourself a sticker

Now – track how you're doing on page 32!

Most and least

 Find the correct sticker to match the number of petals.

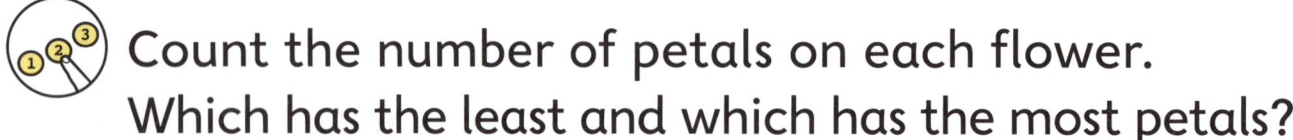 Count the number of petals on each flower. Which has the least and which has the most petals?

 Write the numbers in the boxes.

Least petals ☐   Most petals ☐

Draw two lorries. Give one of them more wheels than the other.

least wheels	**most wheels**

Count how many beads are in each group.

Write the number in the box.

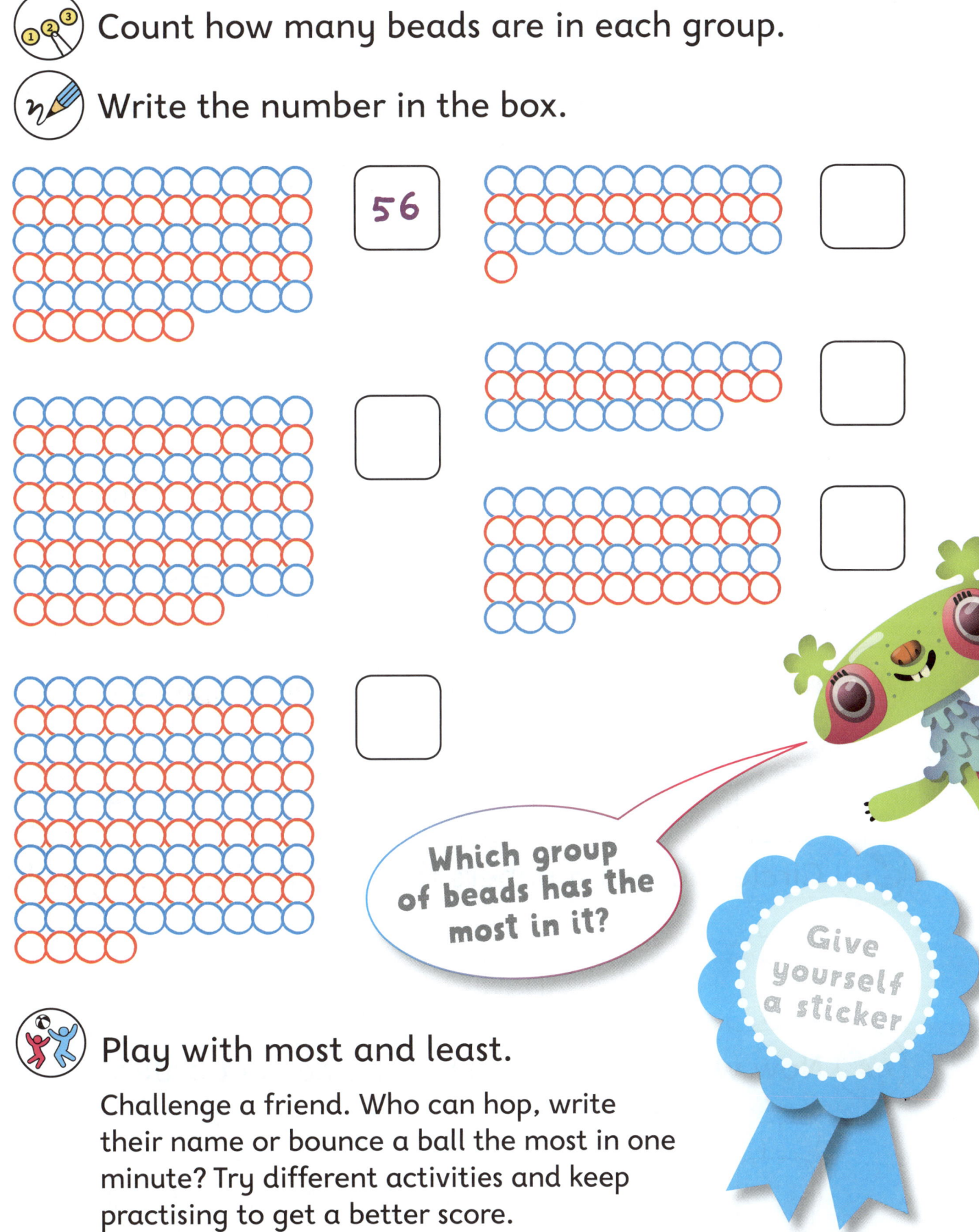

56

Which group
of beads has the
most in it?

Give
yourself
a sticker

Play with most and least.

Challenge a friend. Who can hop, write
their name or bounce a ball the most in one
minute? Try different activities and keep
practising to get a better score.

Now – track how you're doing on page 32!

Ordering numbers

 Circle the two numbers that are in the wrong order.

2   10   17   26   25   31

 Now write the numbers in the correct order.

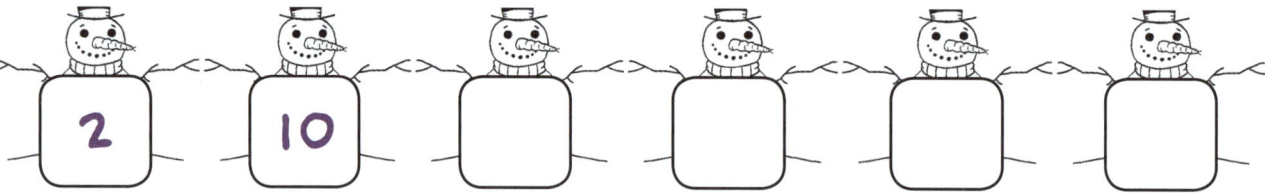

2   10

 Circle the two numbers that are in the wrong order.

44   43   48   57   65   70

 Now write the numbers in the correct order.

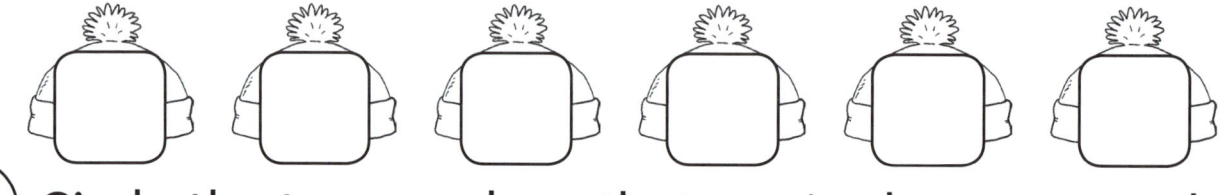

 Circle the two numbers that are in the wrong order.

71   73   79   86   83   94

Now write the numbers in the correct order.

 Write these numbers in order from smallest to largest.

Start with the smallest number first. Cross off each number as you write it on the scarf.

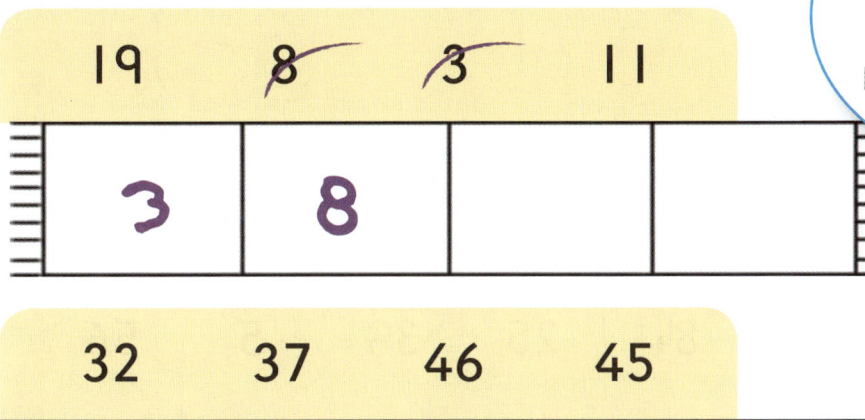

19   8   3   11

| 3 | 8 | | |

32   37   46   45

| | | | |

78   74   61   63

| | | | |

95   80   88   98

| | | | |

 Play with ordering numbers 0–100.

Ask someone to tell you four numbers between 0 and 100. Write each one on a small piece of paper. How quickly can you arrange them in order? Try again with a different set of numbers. Give yourself a challenge by asking for more numbers – five, six, or even seven!

Give yourself a sticker

Now – track how you're doing on page 32!

Number game

Player one stickers

Start 72	97	82	66	84	25	39	5	56	38

55
7
43

Shark race

This is a game for two players.

Choose one shark each and place a counter on each start square. Each player counts 2 spaces and moves their counter around the track in either direction. Compare the numbers you land on.

83	71	12	34	18	91	58	8	27	73

Player two stickers

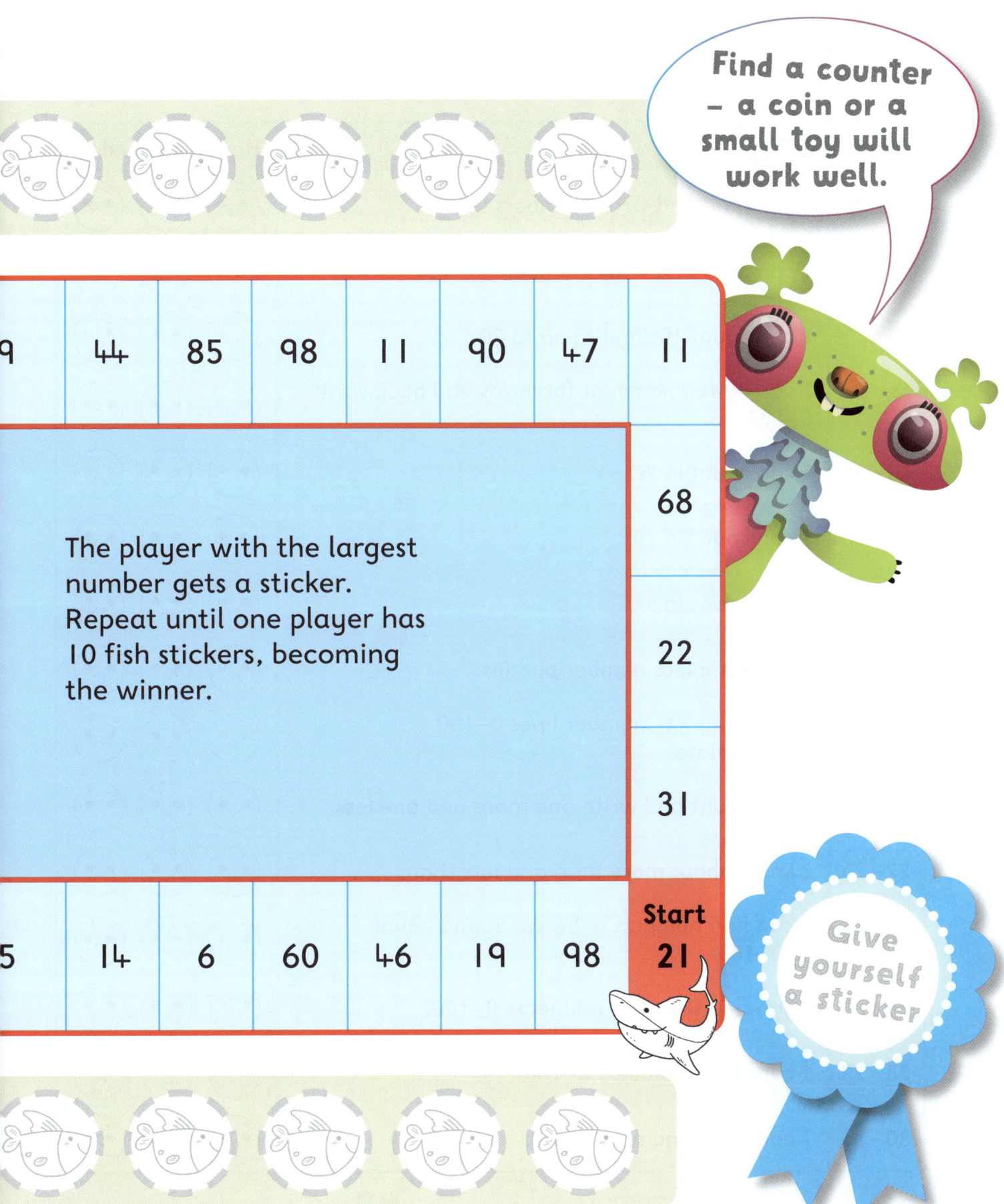

Find a counter – a coin or a small toy will work well.

9 44 85 98 11 90 47 11

68

The player with the largest number gets a sticker. Repeat until one player has 10 fish stickers, becoming the winner.

22

31

Start 21

5 14 6 60 46 19 98

Give yourself a sticker

Now – track how you're doing on page 32!

Progress Chart

Colour in a face.

Page	I Can . . .	How did you do?		
2–3	I can read and write numbers 0–10.	☺	☺	☹
4–5	I can read and write numbers 11–20.	☺	☺	☹
6	I can count 10s and 1s up to 20.	☺	☺	☹
7–9	I can write and count forwards and backwards 0–100.	☺	☺	☹
10–11	I can count in 2s.	☺	☺	☹
12–13	I can count in 10s.	☺	☺	☹
14–15	I can count in 5s.	☺	☺	☹
16–17	I can complete number puzzles.	☺	☺	☹
18–19	I can complete number lines 0–100 and estimate.	☺	☺	☹
20–21	I can count and write one more and one less.	☺	☺	☹
22–23	I know about more than and less than.	☺	☺	☹
24–25	I can find numbers in between and equal to 0–100.	☺	☺	☹
26–27	I can find the most and least 0–100.	☺	☺	☹
28–29	I can order numbers 0–100.	☺	☺	☹
30–31	I can play a number game.	☺	☺	☹

How did YOU do?